Anjara Lalaina Jocelyn RAKOTOARISOA

Quantitative Data Analysis

Anjara Lalaina Jocelyn RAKOTOARISOA

Quantitative Data Analysis

ScienciaScripts

Imprint
Any brand names and product names mentioned in this book are subject to trademark, brand or patent protection and are trademarks or registered trademarks of their respective holders. The use of brand names, product names, common names, trade names, product descriptions etc. even without a particular marking in this work is in no way to be construed to mean that such names may be regarded as unrestricted in respect of trademark and brand protection legislation and could thus be used by anyone.

Cover image: www.ingimage.com

This book is a translation from the original published under ISBN 978-3-8417-9396-6.

Publisher:
Sciencia Scripts
is a trademark of
Dodo Books Indian Ocean Ltd. and OmniScriptum S.R.L publishing group

120 High Road, East Finchley, London, N2 9ED, United Kingdom
Str. Armeneasca 28/1, office 1, Chisinau MD-2012, Republic of Moldova, Europe
Managing Directors: Ieva Konstantinova, Victoria Ursu
info@omniscriptum.com

Printed at: see last page
ISBN: 978-620-8-38938-3

A few introductory words

This handbook is designed to provide students and others with an introduction to the fundamental concepts of quantitative data analysis. Aimed primarily at students who will be carrying out field surveys and collecting data, it is also an introductory guide to sampling methods, without requiring advanced prior knowledge of statistics or mathematics. The manual is designed to be accessible and educational, with practical illustrations to help students understand and assimilate the concepts.

Table of contents

1 Definitions of quantitative data

1.1.Nature of quantitative data

Quantitative data refers to sets of information that can be measured and/or identified (Monjallon, 1980). When information is quantifiable, it can be associated with a specific unit of measurement or categorized according to certain criteria.

Quantitative data can be classified according to their countability. Data is said to be discrete when it is countable, i.e. its number can be counted in a natural way (Moore, McCabe, & Craig, 2017). Mathematically, data is considered discrete when it is represented by integer values. For example, this includes the number of passengers on a plane, the number of students at a university, the number of eggs laid by a hen, or the number of defective parts in a batch.

On the other hand, data is said to be continuous when its value can take on all possible real values, without necessarily being integer or countable. Examples of continuous data include the height of an individual (164.5 cm, 166.0 cm, or 166 cm), the surface area of a plot of land (35.1 m^2, 647.0 km^2, or 12.4 ha), or the capacity of a water tank (78.4 liters, 80.0 decaliters, or 43.4 centiliters).

1.2 Difference between quantitative and qualitative data

There are data that are not necessarily numerical, such as colors (red, blue, white, etc.) or car brands (Toyota, Subaru, Ford, etc.).

In these cases, we speak of qualitative data, which refers to the quality of a given variable.

Qualitative data is generally non-numerical and can be classified as nominal or ordinal. (Tillé, 2010). It is said to be nominal when it represents categories or classes with no defined hierarchy or order. For example, eye color is nominal qualitative data, as no color is considered superior to another. On the other hand, qualitative data is ordinal when it is organized according to a certain order or hierarchy. Examples include military ranks (where general is superior to colonel) or customer satisfaction levels with a product (not satisfied, not very satisfied, moderately satisfied, rather satisfied, very satisfied). So, while quantitative data is generally numerical, qualitative data is non-numerical. The table below illustrates the distinction between the two types of data:

Criteria	Quantitative	Qualitative
Features	Digital measurement	Information describing a non-numeric character
Nature	Digital	Non-digital
Types	- Discrete: countable - Continuous: uncountable	- Nominal: no hierarchy/no order - Ordinal: with hierarchy/with a

Criteria	Quantitative	Qualitative
		certain order of gradation
Examples	- Discrete: number of students at the university - Continue: surface area of a plot of land	- Nominal: marital status (single, widowed, married, etc.) - Ordinal: level of education (primary, secondary, university)

Remarks :

1. In some cases, qualitative data can be quantified. By way of example, colors that are nominal qualitative data can be captured according to their wavelengths. Thus, the color red has a wavelength of around 625 to 740 nanometers (nm), close to 565 to 590 (nm) for yellow and 380 to 430 nm for violet (Villemin, 2017).
2. Qualitative data may (or may not) have a gradation order. In some cases, we can assign it a less obvious type (ordinal or nominal). For example, in the case of the passage of a cyclone, the color variable can have an order scale (green: danger averted, red: danger imminent, etc.). In this illustration, the level of danger is plotted according to a color code, so that in such a situation, color has become a qualitative ordinal datum (each color corresponding to a greater or lesser level of danger).

2 Importance of quantitative data analysis

By transforming raw information into usable quantitative data, data analysis helps us to understand complex phenomena in a variety of situations. This understanding then enables decision-making based on more rational explorations with supporting evidence and foundations.

2.1. Applications in various fields

Quantitative data analysis is used in a wide variety of fields. Here are a few concrete examples to illustrate these cases:

In the social sciences, quantitative data analysis can be used to study social and human behavior. For example, an analysis of data on microfinance in India concluded that access to microcredit increases investment by small businesses. However, this access does not reduce poverty in the short term (Banerjee, Duflo, Glennerster, & Kinnan, 2014).

In behavioral economics, data analysis based on experimental data has suggested that individuals overestimate losses relative to profits. This is the case of loss aversion (Kahneman & Tversky, 1979).

In terms of public health during the covid-19 pandemic, Flaxman et al. (2020) found that, through quantitative analysis of data from Europe, strict containment measures were associated with reduced transmission of covid-19. However, this effectiveness varied according to the speed of implementation and the level of compliance with these measures (Flaxman, 2020). Similarly, in the case of Africa, Eisele et al. (2012) observed that, indeed, the distribution and use of impregnated mosquito nets had significantly reduced child mortality due to malaria (Thomas, 2012).

In the world of marketing, transportation giant Uber has succeeded in maximizing revenues and reducing customer waiting times through the use of quantitative data analysis. Indeed, Uber performed real-time analysis of user demand based on various parameters (climates, festive events, etc.) to fine-tune their fare and was able to balance its supply against demand (Hall & Krueger, 2017).

2.2 Analysis objectives: describe, explore, compare, forecast

Description is one of the goals of quantitative data analysis, exactly as Moore et al. state when they say that "*The goal of data analysis is to provide a clear and accurate description of the data*." (Moore, McCabe, & Craig, 2017) . It consists in the concise and clear summary and presentation of the state/characteristic of a well-defined object or subject. Cochran's (1977) comments on this objective aspect of data analysis that "*data analysis is concerned with the collection, synthesis and interpretation of data to provide insights into underlying phenomena and to guide decision making*" (Cochran, 1977) evokes this descriptive objective. Indeed, by way of example, although French seems to be spoken by a good number of the urban Malagasy population, the data show that it can only be spoken by just under half (47.2%) of this same population. The data therefore show a phenomenon that is not obvious in the sense of the usual perceptions, and relatively non-conforming to common opinion:

		Population totale	Malagasy	Français	Anglais	Autres langues
Milieu de résidence	Urbain	4 608 045	99,9	47,2	19,8	2
	Rural	18 899 925	99,9	17,8	5,4	0,3

Source: MDG - INSTAT - RGPH2018

Within this descriptive framework, quantitative data analysis enables us to calculate numerical indicators that can be used to assess certain characteristics of the object under study.

As well as description, quantitative data analysis also enables us to explore a body of information and extract meaningful features. This exploratory aspect raises the point that "*Data analysis involves the transformation of raw data into meaningful information that can be used to answer research questions and make informed decisions. It involves the application of various statistical methods to summarize, explore and infer information from data.*" (Field, 2013).

By way of example, a comparison between real GDP and carbon dioxide (CO2) emissions into the atmosphere in the case of Madagascar shows that when GDP rises, so does the volume of gas emitted. We speak of a parallel relationship:

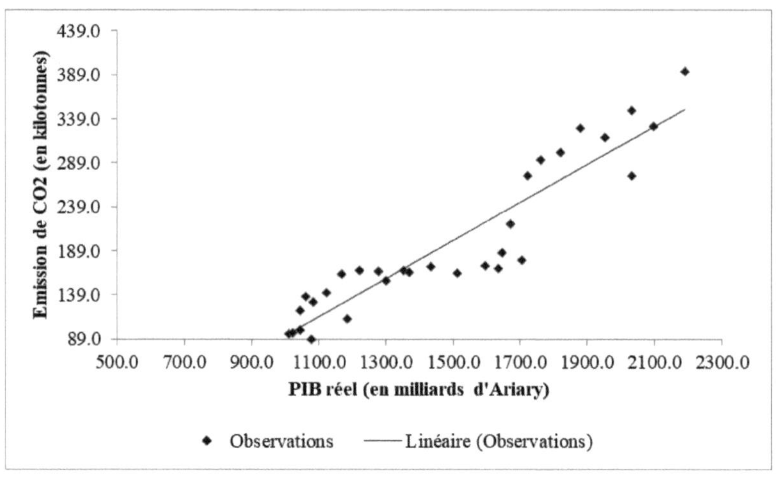

Source: World Development Indicator

Then, quantitative data analysis also offers the possibility of making comparisons by evaluating the difference or possible similarity between two or more distributions of information (Huber & Ronchetti, 2011). For example, it can be used to check whether a treatment is effective by statistically comparing its effect on a treated group with that on an untreated group. This approach is widely used in clinical trials to assess the efficacy of a medical treatment (Huber & Ronchetti, 2011).

Finally, quantitative data analysis provides us with tools for forecasting. This may involve demographic forecasting, assessing the possible impact of an economic policy, or weather forecasting.

PART I: TYPES OF QUANTITATIVE DATA

1.discrete and continuous data

1.1 Definitions and examples

Criteria	Discrete data	Continuous data
Description	Quantitative measurement that can be counted and distinguished	Quantitative measurement that can take any value within a well-defined interval
Features	Accounting and no intermediate value	Measurable with multitudes of possible intermediate values
Value	Whole number	Actual number
Examples	-Population size in a region -Number of words in a text	-Temperature -Water volume in a cistern

1.2 Graphical representations

The graphical representation of a data series depends on the nature of the data, its characteristics and the purpose of the analysis. For example, a curve is often more appropriate for visualizing the evolution of a variable over time. In the example illustrated in the following figure, it is clear to see that the cost of living in Madagascar has generally increased over the years:

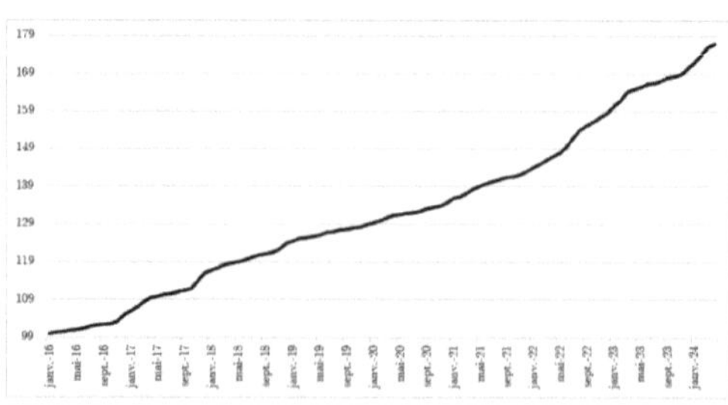

Source: World Development Indicator data graphing

For discrete data evolving over time, the use of bar charts is appropriate, as illustrated in the following figure. Bar charts clearly visualize the variations and trends of a discrete variable over time (Keller, 2018) :

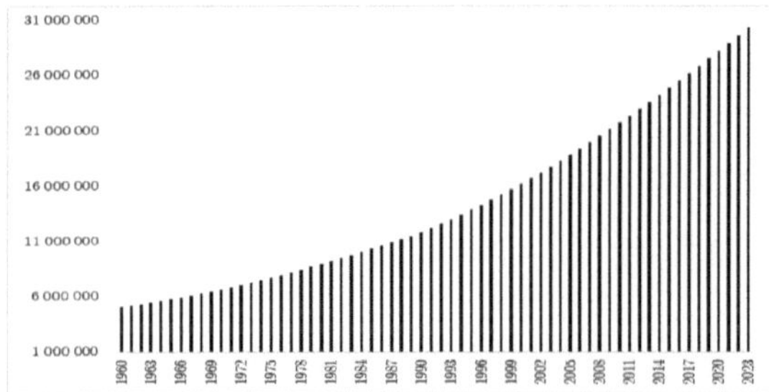

Source: World Development Indicator data graphing

The pie chart is often used to visualize distributions or distributions. It represents a circle divided into sections proportional to the share of data. The total corresponds to 360°. For example, for the distribution of the world's population in 2023, 59.18% live in Asia (see following figure).

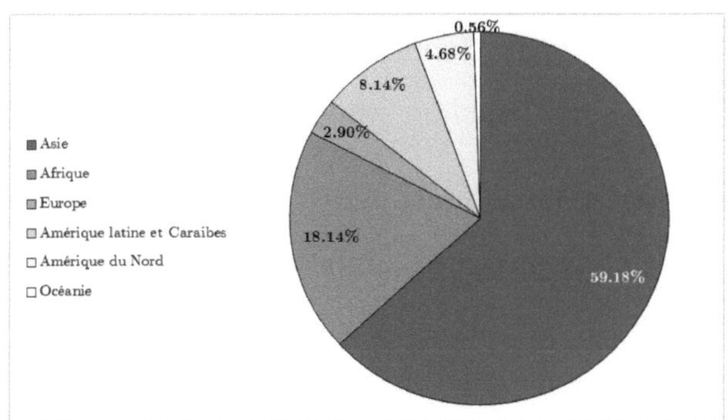

Source: Statista, 2023

2. Measurement scales

Measurement scales are essential for organizing and analyzing the data collected as part of a research project. They determine the type of mathematical and statistical operations that can be applied to the data, as well as how the results can be interpreted (Babbie E. , 2016). There are four types of measurement scales: nominal, ordinal, interval and ratio (Stevens, 1946). Each of these scales has its own specific features, as well as implications for analysis (Burns & Grove, 2010).

2.1. Nominal scale

The nominal scale is the simplest of all measurement scales. It is used to classify data into distinct categories, without order or hierarchy. The data are simply named or labeled, and have no quantitative relationship to each other. (Sullivan, 2010). Values on a nominal scale are qualitative, not numerical. They cannot be meaningfully classified or ordered.

Examples: Gender (male, female), eye color (blue, green, brown), professional category (teacher, engineer, doctor).

In the analysis of nominal data, only operations such as frequency counting or the creation of cross-tabulations are relevant (Trochim, 2006).

2.2. Ordinal scale

Ordinal scales can be used to classify data according to a certain order or hierarchy, but they do not provide information on the gap between categories. It is commonly used in questionnaires where responses are organized in levels or degrees (Sullivan, 2010). Ordinal data indicate an order or ranking. However, differences between scale levels are not necessarily equal or quantifiable (Babbie E., 2016).

Examples: Satisfaction levels (very satisfied, satisfied, not very satisfied, dissatisfied), competition ranking (1st, 2nd, 3rd).

2.3 Interval scale

The interval scale is a numerical scale that not only orders data, but also measures the distance between values. Unlike the ordinal scale, the differences between values on this scale are significant and constant, but there is no absolute zero point (Babbie E. , 2016).

Data on an interval scale can be added and subtracted, but the ratio is meaningless (for example, a temperature of 40°C is not twice as hot as 20°C (Sullivan, 2010).

Examples: Temperature in degrees Celsius, IQ test score.

Statistical methods for the analysis of interval data include mean and standard deviation, as deviations between values are consistent (Trochim, 2006).

2.4. Ratio scale

The ratio scale is the most precise of all measurement scales. Like the interval scale, it has equal intervals between values, but in addition, it has an absolute zero point. This means that all mathematical operations are possible, including ratios (Babbie E. , 2016).

The ratio scale makes it possible to say that one value is twice as large or smaller than another, because there is a zero point that signifies the total absence of the quantity measured (Sullivan, 2010).

Examples: Weight, height, age, income.

Data on a ratio scale allows great flexibility in statistical analyses, including geometric and harmonic means, as well as parametric tests (Trochim, 2006).

Measurement scale	Descriptive statements	Main properties	Examples
Nominal	Categorization or classification without intrinsic order between categories	-No order between categories -No measurable distance between categories	Gender, color
Ordinale	Category classification with gradation	-Category hierarchy Non-measurable distance	Level of satisfaction

Measuremen t scale	Descriptive statements	Main properties	Examples
		between categories	
Interval	Measurement with ordered values		

PART II: QUANTITATIVE DATA COLLECTION TECHNIQUES

1 Collection methods

Data collection is a necessary and unavoidable step in quantitative data analysis. In fact, the method of data collection varies according to the objective pursued (Creswell, 2014). There are three main methods of data collection, namely surveys using questionnaires and/or interviews, experimentation and direct observation (Fowler, 2014).

Each of these methods is adapted to specific contexts. For example, experimentation is much more reliable in the case of clinical trials and in the modeling of time series (or historical or temporal series). Surveys are more favorable in the case of opinion polls (Bryman, 2016).

1.1 Surveys and questionnaires

Surveys and questionnaires are among the most commonly used methods for collecting quantitative and qualitative data in many fields, such as the social sciences, economics, public health and marketing (Fowler, 2014).

Surveys are structured studies that gather information from a population sample of varying size. They can be administered by face-to-face interviews, by telephone, online or by mail. The advantage of surveys is that they can be adapted to include both open (qualitative) and closed (quantitative) questions (Creswell, 2014).

Questionnaires, on the other hand, are standardized measuring instruments containing a series of predefined questions. They allow responses to be standardized to facilitate statistical analysis. Questionnaires can include opinion scales, multiple choice or free response formats (Bryman, 2016).

Example of use: In customer satisfaction surveys, companies use standardized questionnaires to assess aspects such as service quality, customer experience or repurchase intent. The results are then analyzed to identify areas for improvement.

1.2 Experiments

Experimentation is a method of data collection in which the researcher manipulates one or more independent variables in order to observe the effects of this manipulation on one or more dependent variables. It is particularly used in experimental sciences such as psychology, medicine or behavioral economics (Creswell, 2014).

Experimentation can take place in the laboratory, where conditions are controlled, or in the field, where variables are manipulated in a natural environment. Experimental results enable hypotheses to be tested with a high degree of precision and reliability (Bryman, 2016).

Example of use: A marketing company can carry out an experiment to evaluate the impact of different promotions on product sales. By modifying variables such as price or presentation, it can measure the effect of these changes on consumer purchasing behavior (Fowler, 2014).

1.3 Direct observations

Direct observation is a qualitative data collection method that involves observing behaviors, events or situations in their natural environment, without interfering with them. This method is often used in ethnographic studies, behavioral analysis or social science research (Creswell, 2014).

Observations can be structured, i.e. with predefined criteria, or unstructured, and can include video recordings, photos or observation notes. The advantage of this method is that it enables data to be collected in real time and in a contextual manner, often difficult to obtain by other means (Bryman, 2016).

Example of use: As part of a study of customer behavior in a store, a researcher could directly observe how customers interact with products, which departments are most frequented, and how purchasing decisions are made. These direct observations can provide valuable insights into the shopping experience without requiring active intervention from participants (Fowler, 2014).

2.sampling

Sampling is a crucial step in the process of collecting quantitative data. It involves selecting a subset of a population in order to draw generalizable conclusions. The choice of sampling method and sample size influence the quality and reliability of the results obtained (Creswell, 2014).

2.1. Types of sampling

Simple random sampling

Simple random sampling is a method in which the individuals to be surveyed are selected at random. Each member has an equal chance of being included in the sample. So, if we want to survey all the students at the University of Toamasina, we can assign a number to each student and randomly draw numbers to make up the sample.

Stratified sampling

Stratified sampling involves dividing the population into homogeneous sub-groups, called strata, and then sampling randomly within each stratum. This method is useful when the population has varied characteristics and you want to ensure that each sub-group is represented.

For example, in a study of consumer preferences, we can stratify the population by age, gender or income level, then select random samples from each stratum.

Systematic sampling

Systematic sampling involves selecting a sample by choosing a random starting point and selecting every kth member of the population. For example, if you have a list of 1,000 people and you want a sample of 100, you can choose a random starting point between 1 and 10, then select every 10th person on the list.

2.2. Sample size and representativeness

Sample size is a key element in guaranteeing the validity of a study's results. Too small a sample size can lead to unreliable and non-generalizable results, while too large a sample size can lead to wasted resources (Creswell, 2014).

It is crucial to determine an appropriate sample size based on various factors, such as the variability of the population, the desired level of confidence and the precision required in the estimates (Bryman, 2016). An adequate sample size contributes to the representativeness of results and the statistical power of analyses.

A sample is considered representative if it accurately reflects the characteristics of the population as a whole. Appropriate sampling - whether random, stratified or systematic - plays an essential role in ensuring this representativeness (Creswell, 2014).

Good representativeness ensures that the conclusions drawn from the sample can be generalized to the entire population, which is fundamental to the validity of the study results (Bryman, 2016). Consequently, the choice of sampling method must be carefully planned to minimize bias and maximize the reliability of the data collected.

When taking a sample, there are two possible situations. The first is when the size of the population to be surveyed is known and finite. Thus, to find the minimum size required (Israel, 1992)we need a minimum number of observations:

$$n = \frac{\frac{z^2 * p * (1 - p)}{e^2}}{1 + \left(\frac{z^2 * p * (1 - p)}{Ne^2}\right)}$$

With *n* the minimum size required, *z* the z-score coefficient which evaluates the confidence rating (for 95% confidence, z is 1.96 and 2.58 if the desired confidence is 99%), *p* the proportion of the population studied which has the characteristic of interest among the whole, *e* the margin of error retained and N the size of the parent population.

Note: the proportion *p* may not be known a priori. Thus, the value 0.5 (or 50%) is the conservative value that optimizes the error. Consequently, 0.5 is the optimum value for *p*, as it gives the largest sample size, which ensures maximum precision. (Cochran, 1977).

The second case arises when the size of the parent population is unknown or infinite. In such a situation, the minimum sample size (Israel, 1992) to be taken is estimated by :

$$n = \frac{z^2 * p * (1 - p)}{e^2}$$

PART III: DESCRIPTIVE STATISTICS

1.central tendency measurements

1.1.Average

The average is a positional indicator that represents a single value for all the data, as if all the values were identical. For example, if the average age of a population is 25, this means that, in a situation where every individual had the same age, it would be 25.

To calculate the average, we divide the total sum of the data by the number of observations. For example, for the values 5, 4, 9 and 12, the average is calculated as follows: (5 + 4 + 9 + 12) / 4, giving 7.5. Generally speaking, the mean of a series of quantitative data X is defined by this formula:

$$\bar{X} = \frac{1}{n}\sum_{i=1}^{n} X_i$$

With n the number of observations, X_i the value of X for the i-th observation

Example:

Here are the heights (in cm) of the 10 patients in one clinic:

161.1 157.2 147.4 168.6 154.1 169.3 147.3 157.4 159.0 146.3

Since we have 10 patients, the number of observations *n* is then equal to 10. Here, the first observation is 161.1, so X_1 is equal to 161.1, $X_2 = 157.2$, $X_3 = 147.4$, ..., $X_9 = 159.0$ and $X_{10} = 146.3$.

The average patient size is then equal to :

$$\bar{X} = \frac{1}{10}\sum_{i=1}^{10} X_i$$

Or

$$\bar{X} = \frac{161.1 + 157.2 + 147.4 + 168.6 + 154.1 + 169.3 + 147.3 + 157.4 + 159.0 + 146.3}{10}$$

$$\bar{X} = \frac{1567.7}{10} = 156.77$$

The average height is 156.77 cm.

1.2.Median

Before discussing the median, it's essential to understand the concepts of size and frequency. Headcount represents the number of times an observation appears in a data set. To illustrate these concepts, consider the following example:

Let's imagine a village with 25 households, each with a certain number of individuals. Variations in the number of individuals per household illustrate these concepts.

6	10	1	10	8
10	8	1	5	4
8	1	10	4	8
7	2	1	7	9

3	6	3	6	9

In this case, we observe that the number 1 appears 4 times, which means that its number is 4. The number 2 appears only once, so its number is 1. Similarly, the numbers 8 and 10 each have a number of 4. This enables us to draw up the following table of numbers:

X_i	Number $(n)_i$
1	4
2	1
3	2
4	2
5	1
6	3
7	2
8	4
9	2
10	4

Frequency, which expresses the occurrence of a data item in relation to all observations, can be calculated by :

$$f_i = \frac{n_i}{\sum_{i=1}^{n} n_i}$$

We can then deduce the frequency from the table above:

Xi	Number of employees (ni)	Calculation of frequency fi	Frequency (fi)
1	4	4/25	0.16
2	1	1/25	0.04
3	2	2/25	0.08
4	2	2/25	0.08

5	1	1/25	0.04
6	3	3/25	0.12
7	2	2/25	0.08
8	4	4/25	0.16
9	2	2/25	0.08
10	4	4/25	0.16
TOTAL	**25**		**1**

Cumulative frequency is the sum of the frequencies of an observation and those below it. To calculate it, successive frequencies are added together. Here's how to do it:

$$F_i = \sum_{i=1}^{n_i} f_i$$

Xi	Number of employees (ni)	Frequency (fi)	Calculation of cumulative frequency (Fi)	Cumulative frequency (Fi)
1	4	0.16	0.16	0.16
2	1	0.04	0.16+0.04	0.2
3	2	0.08	0.16+0.04+0.08	0.28
4	2	0.08	0.16+0.04+0.08+0.08	0.36
5	1	0.04	0.16+0.04+0.08+0.08+0.04	0.4
6	3	0.12		0.52
7	2	0.08		0.6
8	4	0.16		0.76
9	2	0.08		0.84
10	4	0.16		1
TOTAL	**25**	**1**		

The median is the observation that divides the distribution into two equal proportions of 50% - 50%.

Xi	Cumulative frequency (Fi)	Observation
1	0.16	
2	0.2	
3	0.28	
4	0.36	
5	0.4	The cumulative frequency of 0.5 or (50%) is found between observation 5 and 6.
6	0.52	
7	0.6	
8	0.76	
9	0.84	
10	1	
TOTAL		

To find the median, we proceed by linear interpolation :

5	0.4
Median (Me)	0.5
6	0.52

By linear interpolation :

$$\frac{Me - 5}{6 - 5} = \frac{0.5 - 0.4}{0.52 - 0.4}$$

$$\frac{Me - 5}{1} = \frac{0.1}{0.12}$$

$$\frac{Me - 5}{1} = \frac{0.1}{0.12}$$

$$Me - 5 = 0.8333$$

$$Me = 5.8333$$

50% of households in the village then have a headcount of less than 5.83 and 50% with a headcount greater than 5.83.

1.3.Mode

The mode designates the observation with the highest number of individuals, i.e. the one that appears most frequently. For example, in the series of observations 5, 3, 3, 3, 3, 1, 2, 1, the mode is 3, because it has the highest number of observations, which is 4 (3 appears 4 times).

In our example, we have three modes: 1, 8 and 10. This means that our distribution is multimodal.

Xi	Number of employees (ni)	Frequency (fi)
1	**4**	**0.16**
2	1	0.04
3	2	0.08
4	2	0.08
5	1	0.04
6	3	0.12
7	2	0.08
8	**4**	**0.16**
9	2	0.08

10	4	0.16

2.dispersion measurements

2.1 Scope

The range represents the difference between the maximum and minimum values of a data series. In our example, the range is calculated as follows: 10 (maximum value) - 1 (minimum value) = 9.

$$e = X_{max} - X_{min}$$

2.2 Variance and standard deviation

Variance is a measure of the dispersion of a distribution in relation to the mean. It quantifies the extent to which values deviate from the mean.

$$V(X) = \frac{1}{n}\sum_{i=1}^{n}(X_i - \bar{X})^2$$

Or else

$$V(X) = \frac{1}{n}\sum_{i=1}^{n}X_i^2 - \bar{X}^2$$

	Xi	X²	Xi - Xb	(Xi - Xb)²
	1	1	-4.5	20.25

	2	4	-3.5	12.25
	3	9	-2.5	6.25
	4	16	-1.5	2.25
	5	25	-0.5	0.25
	6	36	0.5	0.25
	7	49	1.5	2.25
	8	64	2.5	6.25
	9	81	3.5	12.25
	10	100	4.5	20.25
TOTAL	**55**	**385**	**0**	**82.5**

$$\bar{X} = \frac{55}{10} = 5.5$$

$$V(X) = \frac{1}{10}385 - 5.5^2 = \frac{385}{10} - (5.5)^2 = 8.25$$

The variance is then 8.25.

The standard deviation is the indicator that quantifies this dispersion through the square root of the variance:

$$\sigma(X) = \sqrt{V(X)}$$

Thus, the standard deviation in the example is 2.87 :

$$\sigma(X) = \sqrt{8.25} = 2.87$$

2.3 Coefficient of variation

The coefficient of variation (CV) is defined as the ratio between the standard deviation and the mean of a data set. A high value of the coefficient of variation indicates a greater dispersion of the data around the mean (Rohatgi & K., 2001).

Generally expressed as a percentage, CV is a unitless measure, making it easier to compare distributions with different measurement scales (Woods & Lichtenfeld, 2018). This property makes it a valuable tool for assessing the relative variability of various data sets.

$$CV = \frac{\sigma(X)}{\bar{X}}$$

For the example, we have :

$$CV = \frac{2.87}{5.5} = 0.5218$$

$$CV = 52.18\ \%$$

PART IV: HYPOTHESIS TESTING

1.null hypothesis and alternative hypothesis

In quantitative data analysis, a hypothesis test is a method used to verify the validity of a hypothesis and draw a conclusion. In this process, the researcher formulates two complementary hypotheses: the first, known as the null hypothesis (H0), represents the default position or initial assumption that we seek to confirm (Cohen, 1988).

This null hypothesis is confronted with a second hypothesis, the alternative hypothesis (H1), which is complementary to H0. The alternative hypothesis expresses the idea that there is an effect or difference, suggesting that the observations are influenced by a particular variable or factor (Field, 2013). This framework is used to determine whether the observed results can be attributed to chance or are statistically significant.

Example of a hypothesis formulation

After the Ambondrona concert at the Coliseum d'Antsonjombe, many Internet users suggested that Ambondrona would be Madagascar's most famous band. To test this "hypothesis", we can carry out a statistical test based on the following assumptions:

H0: Ambondrona is the most famous band in Madagascar (null hypothesis)

H1: Ambondrona is not Madagascar's most famous band (alternative hypothesis)

To test a hypothesis, we use a statistic, noted S, which follows a specific probability distribution, noted L. These two elements make it possible to determine a critical value to which the observed statistic S will be compared. The decision whether or not to reject the null hypothesis depends on this comparison. In general, if the S statistic exceeds the critical value, the null hypothesis H0 is rejected, and the alternative hypothesis H1 is accepted.

2 The chi-square test

The chi-square test is a statistical tool used to verify the existence of an association between categorical variables or to determine whether the distribution of observations corresponds to an expected theoretical distribution, particularly in the case of data independence (Agresti, 2018). This test assesses the interdependence between categorical variables. For example, it can be used to examine whether a child's school enrolment is linked to his or her gender (girl or boy), or to assess the effectiveness of a medical treatment (Siegel & Castellan, 1988). This method is essential in many fields, including social science and public health research.

The chi-square test is performed on a contingency table as follows:

	Y1	**Y2**	**...**	**Yj**	**...**	**Y_{m-1}**	**Ym**	**TOTAL**
X1	A1,1	A1,2	...	A1,i	...	A1,m-1	A1,m	**SL1**
X2	A2,1	A2,2	...	A2,i	...	A2,m-1	A1,m	**SL2**
...	...	...	...	...	...	...	...	**...**

Xi	Ai,1	Ai,2	...	Ai,j	...	Ai,m-1	Ai,m	**SLi**
...	...	...	...	...	...	...	...	**...**
Xn-1	An-1,1	An-1,2	...	An-1,j	...	An-1,m-1	An-1,m	**SLn-1**
Xn	An,1	An,2	...	An,j	...	An-1,m-1	An,m	**SLn**
TOTAL	**SC1**	**SC2**	**...**	**SCj**	**...**	**SCm-1**	**SCm**	**ST**

The principle is to measure the difference between the observations and a situation where the two variables would be totally independent (i.e. unrelated). To simulate this situation of complete independence, we calculate the theoretical numbers using the following formula:

$$E_{i,j} = \frac{SL_i \times SC_j}{ST}$$

	Y1	**Y2**	**...**	**Yj**	**...**	**Ym-1**	**Ym**	**TOTAL**
X1	E1,1	E1,2	...	E1,i	...	E1,m-1	E1,m	**SL1**
X2	E2,1	E2,2	...	E2,i	...	E2,m-1	E1,m	**SL2**
...	...	...	...	...	...	...	...	**...**
Xi	Ei,1	Ei,2	...	Ei,j	...	Ei,m-1	Ei,m	**SLi**
...	...	...	...	...	...	...	...	**...**
Xn-1	En-1.1	En-1.2	...	En-1,j	...	En-1,m-1	En-1,m	**SLn-1**
Xn	In,1	In,2	...	En,j	...	En-1,m-1	En,m	**SLn**
TOTAL	**SC1**	**SC2**	**...**	**SCj**	**...**	**SCm-1**	**SCm**	**ST**

We then evaluate the difference between the observed and theoretical numbers to obtain the independence gap:

	Y1	Y2	...	Yj	...	Ym-1	Ym	TOTAL
X1	E1,1	E1,2	...	E1,i	...	E1,m-1	E1,m	SL1
X2	E2,1	E2,2	...	E2,i	...	E2,m-1	E1,m	SL2
...	...	...	...	...	...	...	...	...
Xi	Ei,1	Ei,2	...	Ei,j	...	Ei,m-1	Ei,m	SLi
...	...	...	...	...	...	...	...	...
Xn-1	En-1.1	En-1.2	...	En-1,j	...	En-1,m-1	En-1,m	SLn-1

Xn	In,1	In,2	...	En,j	...	En-1,m-1	En,m	SLn
TOTAL	SC1	SC2	...	SCj	...	SCm-1	SCm	ST

The deviations from independence are :

$$D_{i,j} = A(i,j) - E(i,j)$$

	Y1	Y2	...	Yj	...	Ym-1	Ym	TOTAL
X1	D1,1	D1,2	...	D1,i	...	D1,m-1	D1,m	0
X2	D2,1	D2,2	...	D2,i	...	D2,m-1	D1,m	0
...	...	...	...	...	...	...	...	0
Xi	Di,1	Di,2	...	Di,j	...	Di,m-1	Di,m	0
...	...	...	...	...	...	...	...	0
Xn-1	Dn-1,1	Dn-1,2	...	Dn-1,j	...	Dn-1,m-1	Dn-1,m	0
Xn	Dn,1	Dn,2	...	Dn,j	...	Dn-1,m-1	Dn,m	0
TOTAL	0	0	0	0	0	0	0	0

The chi-square is obtained from :

$$\chi^2 = \sum_{i=1}^{n} \sum_{j=1}^{m} \frac{D^2(i,j)}{E^2(i,j)}$$

H0: X and Y are independent (no relationship)

H1: X and Y are related

The null hypothesis H0 is rejected at α margin of error if χ^2 greater than K^2. Recall that K^2 is the threshold chi-square for α margin of error at ν degree of freedom (chi-square table).

The margin of error α is exogenous (to be set a priori). The degree of freedom is calculated by :

$$\nu = (n - 1)(m - 1)$$

With n the number of rows and m the number of columns.

Example:

A medical laboratory is conducting a clinical trial to determine whether a treatment is effective in curing a disease. To this end, it has collected data from 100 patients, which are presented in the following contingency table:

	Did not receive treatment	Having received the treatment	TOTAL
Healed	11	47	58
Not cured	39	3	42
TOTAL	50	50	100

Theoretical headcount

	Did not receive treatment	Having received the treatment	TOTAL
Healed	29	29	58
Not cured	21	21	42
TOTAL	50	50	100

Departures from independence

	Did not receive treatment	Having received the treatment	TOTAL
Healed	-18	18	0

Not cured	18	-18	0
TOTAL	0	0	0

χ^2 calculation

	Did not receive treatment	Having received the treatment	TOTAL
Healed	11.17241379	11.17241379	22.344828
Not cured	15.42857143	15.42857143	30.857143
TOTAL	26.60098522	26.60098522	53.20197

We then have $\chi^2 = 53.20197$.

At 5% (or 0.05) risk of error and $v = (2-1)(2-1) = 1$ degree of freedom, the critical chi-square K^2 (according to the chi-square table) is 3.84 (see how to read the following table):

Risque/marge d'erreur

Degré de liberté

α / v	0,99	0,975	0,95	0,90	0,10	0,05	0,025	0,01	0,001
1	0,0002	0,001	0,004	0,016	2,71	**3,84**	5,02	**6,63**	10,83
2	0,02	0,05	0,10	0,21	4,61	**5,99**	7,38	**9,21**	13,82
3	0,11	0,22	0,35	0,58	6,25	**7,81**	9,35	**11,34**	16,27
4	0,30	0,48	0,71	1,06	7,78	**9,49**	11,14	**13,28**	18,47
5	0,55	0,83	1,15	1,61	9,24	**11,07**	12,83	**15,09**	20,51
6	0,87	1,24	1,64	2,20	10,64	**12,59**	14,45	**16,81**	22,46
7	1,24	1,69	2,17	2,83	12,02	**14,07**	16,01	**18,48**	24,32
8	1,65	2,18	2,73	3,49	13,36	**15,51**	17,53	**20,09**	26,12
9	2,09	2,70	3,33	4,17	14,68	**16,92**	19,02	**21,67**	27,88
10	2,56	3,25	3,94	4,87	15,99	**18,31**	20,48	**23,21**	29,59

Chi-square table

TABLE DU χ^2

La table donne la probabilité α pour que χ^2 égale ou dépasse une valeur donnée, en fonction du nombre de degrés de liberté v. Exemple : avec v = 3, pour χ^2 = 0,11 la probabilité α = 0,99.

α / v	0,99	0,975	0,95	0,90	0,10	0,05	0,025	0,01	0,001
1	0,0002	0,001	0,004	0,016	2,71	**3,84**	5,02	**6,63**	10,83
2	0,02	0,05	0,10	0,21	4,61	**5,99**	7,38	**9,21**	13,82
3	0,11	0,22	0,35	0,58	6,25	**7,81**	9,35	**11,34**	16,27
4	0,30	0,48	0,71	1,06	7,78	**9,49**	11,14	**13,28**	18,47
5	0,55	0,83	1,15	1,61	9,24	**11,07**	12,83	**15,09**	20,51
6	0,87	1,24	1,64	2,20	10,64	**12,59**	14,45	**16,81**	22,46
7	1,24	1,69	2,17	2,83	12,02	**14,07**	16,01	**18,48**	24,32
8	1,65	2,18	2,73	3,49	13,36	**15,51**	17,53	**20,09**	26,12
9	2,09	2,70	3,33	4,17	14,68	**16,92**	19,02	**21,67**	27,88
10	2,56	3,25	3,94	4,87	15,99	**18,31**	20,48	**23,21**	29,59
11	3,05	3,82	4,57	5,58	17,28	**19,68**	21,92	**24,73**	31,26
12	3,57	4,40	5,23	6,30	18,55	**21,03**	23,34	**26,22**	32,91
13	4,11	5,01	5,89	7,04	19,81	**22,36**	24,74	**27,69**	34,53
14	4,66	5,63	6,57	7,79	21,06	**23,68**	26,12	**29,14**	36,12
15	5,23	6,26	7,26	8,55	22,31	**25,00**	27,49	**30,58**	37,70
16	5,81	6,91	7,96	9,31	23,54	**26,30**	28,85	**32,00**	39,25
17	6,41	7,56	8,67	10,09	24,77	**27,59**	30,19	**33,41**	40,79
18	7,01	8,23	9,39	10,86	25,99	**28,87**	31,53	**34,81**	42,31
19	7,63	8,91	10,12	11,65	27,20	**30,14**	32,85	**36,19**	43,82
20	8,26	9,59	10,85	12,44	28,41	**31,41**	34,17	**37,57**	45,31
21	8,90	10,28	11,59	13,24	29,62	**32,67**	35,48	**38,93**	46,80
22	9,54	10,98	12,34	14,04	30,81	**33,92**	36,78	**40,29**	48,27
23	10,20	11,69	13,09	14,85	32,01	**35,17**	38,08	**41,64**	49,73
24	10,86	12,40	13,85	15,66	33,20	**36,42**	39,36	**42,98**	51,18
25	11,52	13,12	14,61	16,47	34,38	**37,65**	40,65	**44,31**	52,62
26	12,20	13,84	15,38	17,29	35,56	**38,89**	41,92	**45,64**	54,05
27	12,88	14,57	16,15	18,11	36,74	**40,11**	43,19	**46,96**	55,48
28	13,56	15,31	16,93	18,94	37,92	**41,34**	44,46	**48,28**	56,89
29	14,26	16,05	17,71	19,77	39,09	**42,56**	45,72	**49,59**	58,30
30	14,95	16,79	18,49	20,60	40,26	**43,77**	46,98	**50,89**	59,70

Source : http://www.leblogdutesteur.fr/test-de-khi-deux/

Bibliography

Agresti, A. (2018). *Statistical Inference.* Duxbury Press.

Babbie, E. (2016). *The Practice of Social Research.* Cengage Learning.

Babbie, E. R. (2016). *The Practice of Social Research.* Cengage Learning.

Banerjee, A., Duflo, E., Glennerster, R., & Kinnan, C. (2014). *The Miracle of Microfinance? Evidence from a Randomized Evaluation.* American Economic Journal: Applied Economics.

Bryman, A. (2016). *Social Research Methods (5th ed.).* Oxford University Press.

Burns, N., & Grove, S. K. (2010). *The Practice of Nursing Research: Appraisal, Synthesis, and Generation of Evidence (6th ed.).* Elsevier.

Cochran, W. (1977). *Sampling Techniques (3rd ed.).* New York: John Wiley & Sons.

Cohen, J. (1988). *Statistical Power Analysis for the Behavioral Sciences (2nd ed.).* Lawrence Erlbaum Associates.

Creswell, J. W. (2014). *Research Design: Qualitative, Quantitative, and Mixed Methods Approaches (4th ed.).* Sage Publications.

Field, A. (2013). *Discovering Statistics Using IBM SPSS Statistics (4th ed.).* Sage Publications.

Flaxman, S. (2020). *Estimating the effects of non-pharmaceutical interventions on COVID-19 in Europe.* Nature.

Fowler, F. J. (2014). *Survey Research Methods.* Sage Publications.

Hall, J., & Krueger, A. (2017). *An Analysis of the Labor Market for Uber's Driver-Partners in the United States.* ILR Review.

Huber, P., & Ronchetti, E. (2011). *Robust Statistics.* John Wiley & Sons.

Israel, G. (1992). *Determining Sample Size.* University of Florida IFAS Extension.

Kahneman, D., & Tversky, A. (1979). *Prospect theory: an analysis of decision under risk.* Econometrica.

Keller, G. (2018). *Introduction to Statistics.* Duxbury Press.

Krejcie, R., & Morgan, D. (1970). *Determining sample size for research activities.* Educational and Psychological Measurement.

Monjallon, A. (1980). *Introduction à la méthode statistique.* Paris: Vuibert.

Moore, D., McCabe, G., & Craig, B. (2017). *Introduction to the Practice of Statistics (9th ed.).* New York: W. H. Freeman.

Rohatgi, V. K., & K., S. A. (2001). *An Introduction to Probability and Statistics.* Wiley.

Siegel, S., & Castellan, N. J. (1988). *Nonparametric Statistics for the Behavioral Sciences (2nd ed.).* McGraw-Hill.

Stevens, S. S. (1946). *On the Theory of Scales of Measurement.* Science.

Sullivan, L. M. (2010). *Statistics in Medicine.* Wiley-Blackwell.

Thomas, E. (2012). *Malaria prevention in infants and children in sub-Saharan Africa: A systematic review and meta-analysis.* Lancet Infectious Diseases.

Tillé, Y. (2010). *Resumé du Cours de Statistique Descriptive.* University of Neuchâtel.

Trochim, W. M. (2006). *The Research Methods Knowledge Base (2nd ed.).* Cengage Learning.

Villemin, G. (2017, August 13). *Electromagnetic waves.* Retrieved from Lumvisib: http://villemin.gerard.free.fr/Science/Lumvisib.htm

Woods, T. M., & Lichtenfeld, M. (2018). *Statistical Methods for the Social Sciences.* Pearson.

Printed by Books on Demand GmbH, Norderstedt / Germany